AuthorHouse™
1663 Liberty Drive
Bloomington, IN 47403
www.authorhouse.com
Phone: 1 (800) 839-8640

Scripture quotations marked NKJV are taken from the New King James Version.
Copyright © 1982 by Thomas Nelson, Inc. Used by permission. All rights reserved.

Published by AuthorHouse 12/10/2015

ISBN: 978-1-5049-5966-7 (sc)
978-1-5049-5967-4 (e)

Print information available on the last page.

Any people depicted in stock imagery provided by Thinkstock are models,
and such images are being used for illustrative purposes only.
Certain stock imagery © Thinkstock.

This book is printed on acid-free paper.

authorHOUSE®

The Case For God

A Math Book

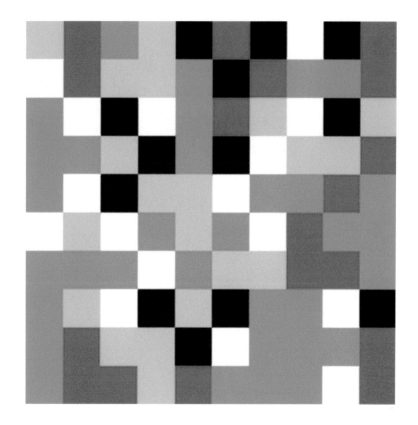

Clancy John Imislund

Table of Contents

Prologue

Hello. Let us start this easy. My name is Clancy and I am a computer engineer with a great amount of exposure to technology and science. That being said, I personally believe in God or whatever name you would like to use and the following pages will go through that process. This text is not meant to sway you either way and no religion will be enforced on you. It is just some information that tries to marry Science and Creation. There is no "Thou shalt not…" here. I will also try to put doubt into the minds of both the religious zealots and those that claim that no God was required to create this universe.

Contained in these pages, there will be a few puzzles and questions. They may be unnerving for both atheists and the believers in God but that is the purpose. Everything here is based on pure logical observation and not on any mythical or religious foundation. There will be a few lines of tedious math and Bible verse but those you may choose to skip over. They are only there for your perusal and to help make points.

At the end of each chapter, there will be a question. Hopefully they are all good enough to make it into your meditation schedule. They sure made it into mine!

Anyway, I hate long and boring prologues as much as the next person. As your attorney inside of these book covers, I advise we commence!

Chapter 1

"In the Beginning"

Okay. Here is an old joke. It is always good to break the ice. It will get much nastier after this point so let's laugh it up for now.

A Fella: "God?"

God: "Yes, my child?"

A Fella: "God? How long is a billion years to you?"

God: "It is less than a second, my child."

A Fella: "God? How much is a million dollars to you?"

God: "It is less than a penny, my child."

A Fella: "God? May I have a penny?"

God: "Yes, my child--in a second."

That strange joke is germane for this first chapter.

Now I will proceed and try to get the boring stuff out of the way quickly. I stayed away from all of the math I could so as not to bore any of us.

It will be presented by way of a fictional court trial. This court does not have the rules of a conventional courtroom so I have omitted the location of this arena. The "Three Stooges" would be better counselors. My friends and I can speak more freely here though.

Enter it if ye dare...

Black Holes
(This is a section you can skip if you understand them already)

Before this first part, let us have a very short primer regarding "Black Holes". They are collections of matter so dense that not even light can escape their gravitational grip. Nothing that we know of may return after it crosses what is called the "Event Horizon". They are without dimension and they cannot be measured except by their gravitational effects on nearby objects.

The word for such a thing is a "Singularity". It only takes a chunk of matter about 3 times more massive than that of our sun to create one after the Hydrogen is expended. They bend space and time as Einstein has already theorized and proven through many experiments conducted by folks even smarter than him. An exercise for the reader says me.

Okay...

There are four fundamental forces in the universe (That we can detect, at least). They are the Weak Nuclear force, the Strong Nuclear force, the Electromagnetic force, and finally, our pal, Gravity. The first three forces are those that keep you from falling through the floor or walking through walls. When you stub your toe at night or you punch out the jerk that hangs out on the corner, the atoms in your body are never really touching anything. Those forces are so strong that contact is never made and they defeat Gravity -- to a point.

When you were a kid, did you ever try to touch two positive or negative ends of a magnet together and have them shove each other away? You could force them together with the smaller versions but with larger magnets, then tough luck. It is the same thing except that the fundamental forces operate on a much smaller scale and they are much more powerful. Gravity overrides them all when a sufficient amount of matter exists in a tight sphere. We are all falling constantly but those forces are stopping us from becoming part of the Earth's core.

Here is the first boring math thing. It is not complete but it was good enough to get men on the moon. Albert Einstein used this extensively and expounded on it. For this, we don't care. We have enough to define a Black hole.

$$F_g = -G\frac{m_1 m_2}{r^2}$$

This is the Force of Gravity defined mathematically. G is the gravitational constant. The subscripted m values are the masses of the two bodies interacting, and r is the distance between their centers of mass. When r approaches zero, this function generates an asymptote for which the first three forces yield. This is an inverse square and the value also negated. It collects all mass together very quickly and compactly.

When r approaches zero then yes, you can fall through floors and punch right through that fella on the corner if you can make it there in time. Gravity wins and that is how a Black Hole operates.

It turns into nothing but inescapable gravity and an immeasurable singularity except through its effect on other bodies. It takes up virtually no space at all--at least nothing for which you can assign dimensions.

Boring math over for now and armed, let's get back to the court. ..

The Bailiff:

All Rise (The Judge enters the room)

The Judge:

Counsel for the Defense, you may now begin your argument.

The Atheist:

I object, your Honor. This is not a science lesson and it certainly has nothing to do with his "Proof of God".

The Judge:

He never said anything about a "Proof of God" and it is not the subject of this case. He must be allowed his chance. I don't know what that physics report he sent to us was all about and I will watch him closely. Objection overruled. Now please begin, Counselor.

Creation vs. Evolution

The Book of Genesis starts with:

"In the beginning, God created the heavens and the earth. The earth was without form, and void.".

Me:

"The earth was without form, and void."-- Now that sure sounds like the perfect description of a singularity. As mentioned before, it only takes a mass of about 3 times larger than that of our sun to create one. We are now talking about all of the mass that was the early universe and packed into a singularity. Nothing could escape. All of the mass in the universe is several trillion times more massive than that of our sun and every scientist will concur. I will let my atheist friend have the floor now...

The Atheist:

Science has proven that a singularity containing all of the matter in the universe seemed to appear from nowhere about 13.5 billion years ago. We do not know how or why but we are on it. We can trace it back to its original point in space by studying red-shifts in background radiation that we have measured. It did not happen in a day.

It happened in a period of time that is too short for us to measure or comprehend because of the effects given by Quantum Mechanics.

(A day. That is 24 hours or so to us earthlings if you want to take it literally. We'll get to that shortly.).

Science has debunked his introduction of the universe. If you look at the physics, it is not possible for this to have happened in a day. I do not expect you, the jury, to understand. I just want you to know that everything you learned before was a fallacy from a storybook called "The Bible". Your turn, counselor...

Me:

Okay, My turn...

The Book of Genesis goes on with:

"Then God said, 'Let there be light'; and there was light."

Me:

This means the object that was *"without form, and void."* violated every mathematical law of Physics with new laws that even Einstein could not fathom(Super-massive Black Holes don't just explode and there is no science to prove otherwise). The universe just became "magically", in your opinion. Your turn again, counselor...

The Atheist:

There was an event. We call it 'The Big Bang'. This mysterious singularity erupted at the speed of light and probably faster than that. We are not sure how or why. We can measure the residue though and we have several hundred books you can read to understand our observations if you have a strong enough mind. They are complicated. As I said before, we are on it.. Please be patient.

Yes. We have compiled much over several millennia. It is complex and certainly not for children. The chapters on red-shift, background radiation, and microwaves reveal the true clues to the advent of the universe. Argue that, counselor.

The Judge:

People, let us take a break now and commence in 30 minutes.

Me:

Now wait a minute, your Honor. I have a few things I would like to add. If you go through the first chapter of Genesis, it unfolds into the story of Evolution almost perfectly. The Universe, Earth, plants, and animals all came before as Evolutionary theories have practically proven. Man came on the last "day".

We will get to the "day" in a moment but let's focus on Genesis and Evolution first. The "Woman came from Adam's rib" bit fits too. Most living things are asexual even to this day as was most every living thing from prehistory.

Somewhere along the line, certain animals evolved opposite sexes. Genesis says this in just a few short lines and it is extremely concise and clear -- not to mention, poetic. There is no need for those endless centuries of study and textbooks. It is simple enough for a child to understand. The Bible is aimed at children both young and old.

The Judge:

What does that have to do with anything in this case?

Me:

In Germany, a proper greeting is "Hallo". If you go to a Spanish speaking country, you would say "Hola". If you went to a Russian speaking land, you would say "привет". Those are completely different words, your Honor.

The Judge:

Again. What does that have to do with anything? Stay on point. You are a moment away from being held in contempt of this court.

Me:

I am on point, your Honor. The short language lesson was there to show that words spoken differently can have exactly the same meaning. If this is the discrepancy between "Evolution" and "Creation" then we are wasting our time in this courtroom today. Wouldn't you agree there, judgey-poo?

The Judge says:

Watch it! You are on thin ice! This is quite serious and you will address me as 'Your Honor'!!!

Me:

    Okay and sorry about that. All I am saying is that the stories of "Creation" and "Evolution" are identical; They are merely written in different languages but the chronology is in-step. People fight and kill each other over this as we are today in this weird room. Anyway, that is all I have to say about that. I am going to get a sandwich now as per your request for a break. Aloha, Judgilicious. See ya at the noodle factory!

The Judge: "BAILIFF! CATCH HIM!"

The Day

Two weeks later and still on the run from this judge. Funny thing how you can get used to eating sardines you cooked in a makeshift fire-pit under the pier. We can talk about a "day" now though. They won't catch me!

There are hardcore Bible enthusiasts that will tell you that "A day is a day is a DAY". Light and dark. The universe and everything was created during 6 periods of this time--Thump on the Bible and believe. There are atheists that will tell you that there is no way certain events could've occurred in this period. We can prove it to you both mathematically and logically. Oh Boy! Let's dig in!

The Earth rotates around a star-- our Sun. The Earth's axis is not in alignment with that of this star. It wobbles and depending on the location you are occupying, the periods of day and night vary. There also another thing called "precession" which causes the axis to be tampered with even more. The average period between night and day is roughly 24 hours on this planet.

It orbits the star and takes 364 or 365 of day/night periods for a full revolution around. I can get into an argument with my odd friend in Lapland during Winter and say to him "Sunrise is at 6:40am". He will say "NO! It is at 11:00am.". Get out the Pugil sticks! This could take all night (Hopefully my night. No way could I hang in there for 20 hours with Olaf.). That part is explained. Let's move on with the "day"...

There are 8 other planets revolving around our tiny star. Let us talk about a few of them. They participate in all of the same activities described above. I will translate this into earth days for perspective.

- The first planet, Mercury, has a day that is the length of around 59 Earth days
- The second planet, Venus, has a day that is the length of around 116 Earth days
- The third planet is Earth. We already went over that.
- The fourth planet is Mars. A day there is roughly equal to an Earth day.
- The fifth planet is Jupiter. The day there is around 10 hours of Earth time.

There are four more planets but this is getting extremely boring!

There are nine planets (Yes--I am including the recently-demoted Pluto) that orbit around our star. There are trillions and trillions of other planets orbiting more trillions of other stars in this universe. Astronomers and astrophysicists find more of them every day. Each has a different length of day/night period. There are many definitions and notions we have invented to explain the different passages of time--"Epoch, Eon, Period, Second…etc". There are hundreds of thousands more but I will not list them as I am not trying to bore you and that "King of the Hill" judge who is after me is right on my tail. No time to tarry so I will leave it there.

The Year

The Bible states that our planet has been here for around 5000 years but the astrophysicists claim a period of about 4.5 billion years. That is a pretty big disagreement!

We already talked about the differences in the length of a day on the many trillions of planets in the universe. They each have a different length for a period called a "Year" too-- even within our own small solar system. I will leave it at that. All scientists and atheists will agree with us on this point.

We are rolling up to our first real question in this part. My personal belief is that "Day", "Night", and "Year" are metaphors for "Beginnings" and "Endings" and not definitions of earthly quanta we judge by sunrise and sunset. The Bible often speaks like that and as was said, The Bible was written for children of all ages. The difference is that the book was written 3000 or so *earth* years ago and those folks did not have a "Hubble Space Telescope" and the myriad of other detectors that are available now.

I say the book of Genesis clarifies via allegory--open and shut. No reason to quibble about the length of time elapsed. It does not matter--only the chronology is what needs to match up for us to agree with our measurements.

Question

"If we cannot agree on the terms related to the passage of time on our small planet or the rest of the Universe, how can we presume to know the length of a day or a year for God?".

Holy Mackerel! There is that judge and her goons again down by the market! I gotta run! See ya in the next chapter!

Chapter 2

Evolution

GREAT NEWS! The Judge and I got together and worked out our differences. I apologized and delivered to her a fancy gift basket. It had various fruit, muffins, cheese, crackers, and wine. She loved it so I only received a $100 fine and a 6-month probationary community-service plan.

She gave the muffins to the goons and they have since stopped their surveillance. There was a single delicious pear in that basket. The Judge re-gifted and had it delivered to my atheist friend. Unlike the apple presented to Adam and Eve, he is very welcome to eat it. The doctors and the orderlies over at the rehab clinic say he must wait the mandatory ten days before he can have it though. Be well, friend! Our court case can commence in two weeks.

In the last chapter, we touched briefly on the Theory of Evolution. I wholeheartedly believe it to be true and it is a shame that Charles Darwin and his ilk had to jump through so many hoops of condemnation to get it out there.

The "Index Librorum Prohibitorum" was where it was catalogued by the Church. There are many still in disbelief regarding those works so we are going to put it under the microscope here. Yes. It is another argument for God. There are a few boring concepts that we have to get through first so let us get it over with. The next one will really give my friend a headache...

Entropy
(Skip this if you already know it)

This word defines an energetic system that constantly tends towards chaos. In the physical laws of Thermodynamics, it is the "Murphy's Law". Everything, no matter how well ordered, will proceed into a state of disorder. There is much proven math to sustain this. I will skirt around this math as it is really not needed for our Evolution microscope study. It is just the concept of "Entropy always increases." that is important.

Einstein's equation:

$$E=mc^2$$

…applies to us but we will not get into Relativity or any of that stuff (It would definitely make our definitions of "Day" and "Year" more complicated though!). The crucial thing is that the equality of matter and energy means that Entropy affects all things, living and inanimate.

Science and the Bible both agree that all life came from some asexual primordial soup and that we and everything moved forward from there. They just have different ways of expressing it. We've already discussed that but it is valuable that we understand this before we bandy words like "Evolution" and "Entropy" around. We will also touch on Genetics here. Before we get to that, let us start with a story about a moth.

The Moth

In England, there was a moth. This moth was white in color and it blended in perfectly with the white bark of the trees in its habitat. It confounded the birds that preyed upon it by using a perfect camouflage. Now this moth also had a strain that was black in color. That version was easy pickings for those birds. It was very rare to see one as most of them ended in a bird's stomach.

Around 1800 or thereabouts, England had its Industrial Revolution. This was great for the country and economy of that land. Many useful products were created in large factories that could be employed for domestic or foreign usage. The country was alive and booming with new jobs and opportunities. Life became much easier!

Not so for our friend, the moth. Many of these factories burned coal to drive their contraptions. It is a minimal source of energy but good enough to drive a steam engine. It works. The side effect of burning coal is that large amounts of dark smoke and residue are emitted into the atmosphere and environment. This residue covered most everything and for our moth-- the factory exhaust made the white trees turn black. The birds had a field day and the white moth virtually disappeared. The black strain thrived however.

This is a perfect example of "Natural Selection" in action and Charles Darwin on his ship, "The Beagle", never needed to set sail. It was right in front of him in his native land and it was easy enough for a child to understand.

What Came First--The Chicken or the Egg?

Before we get into Genetics and then the final point, I must answer this question finally and forevermore. The creature that put forth that egg was most likely a bird much like a chicken but it was not a chicken. The chicken that emerged from that egg was no longer able to reproduce with the species that had created it. How did this happen? It was a mutation of some kind but I sure don't know. I will lay on science for this and get to it in the next part before all of what you have up there is tied together. Anyway, the answer is "The Egg".

I am happy that Judge is not here as the gavel would be broken by now and I would probably have to pay for it plus punitive damages. Okay. Just one last thing and we will put the whole shebang together. I know it is not making sense so far!

Genetics

Every living thing has in it contained a long sequence of microscopic chemical instructions. Even "germs" are included in this group. These instructions guide everything in an organism's form and structure and the order is important. We call it "DNA". It is a very complex string composed from 4 molecular building-blocks linked together in highly organized strands(There is a 5th building-block but that is only used when the asexual cell is trying to reproduce). Every cell in every living thing has these including the guy on the corner that needs the beating.

If the genetic sequence is altered during formation, your child may come out as a chimpanzee or a squirrel. These alterations come in many forms and it is why I cited the chicken. The many forms of mutation are not important. It is just that we must agree that it occurs. There is much math and chemistry to describe it but for this, we don't care. What we care about is the order and operation of these systems. They violate the law of Entropy and that cannot be denied.

All Together Now

There were many seemingly random and boring topics with moths and chickens described up there. I, for sure, nearly fell asleep reading them! Now let us tie them together and ask a question...

Question

"If the proven laws of Thermodynamics, including Entropy, are correct, then what is the force behind all of the order we see in ourselves, in nature, and the rest of the universe? Magic?"

It does not have to be a "God" but something is driving it and disobeying all observable and measurable physical laws. That is not my whole case but it is something to consider. At least it does not disprove a God and I will get to that when back at court in the morning with the "Transistor" section. One more rest-stop before that though. On we go!

Chapter 3

More Bible

Those two previous chapters were extremely tedious and dull so let's try to spice it up a little. I need to get my creative juices flowing as I must report to the court tomorrow to continue the showdown with that Atheist and a very impatient Judge. The Atheist is doing much better now and is ready to go in the morning. The pear did the trick. This is excellent to hear! I never wished any hurt on him.

What we will talk about in this chapter are a few stories and try to explain their purposes. These are fun stories but they will soon become quite offensive to some. As far-fetched as it sounds, this is a defense for the Bible and it all does fit into this case proper.

We will start with a simple one...

The Lorax

When I was a child, I loved to read the works of Dr. Seuss. One of my favorites was the story of "The Lorax". It is an interesting tale that was centered around an idyllic village with meadows and forests of "Truffula trees". There were many happy creatures living there called "Humming Fish" and "Swomie Swans". An odd creature also lived there but it does not appear yet.

The story begins with a child entering this village and all of the trees, fish, and swans were gone. It was a desolate, flat, and gray landscape now. The child was confused. He found the last resident of this village, the person responsible (Seuss named him 'The Once-ler'), smoking a cigar from a second-story window. The child was made to pay a small tithe to hear the story from this person and this he did.

The Once-ler told of a time when the land was green and abundant. The swans, fish, and trees were bountiful. He also knew it for certain that he could use these things to start a business. This business would be involving the harvest of those trees and inventing a thing called a "Thneed"--"Something, something everyone needs!". In this process, he had awoken a creature called "The Lorax" that emerged from the stump of a dead tree. Using the illustrations, it was part Beaver and part Manatee in appearance and it stood upright about 3 feet tall. For this, we don't care.

The Lorax spoke to the people and said "I am the Lorax and I speak for the trees as the trees have no tongue.". He came from his sleep and warned this village of the huge mistake they were making but the townsfolk did not care as they were making a bazillion, gazillion, kadllion druloos from this(<= Seuss talk!). The Lorax nagged and nagged but nobody cared or listened.

After many years, there was the day when the last Truffula tree was finally cut down. The Lorax made a sad look, grabbed the seat of his pants, and flew off into the distance--never to be seen again.

The Once-ler finished his story and then he dropped down the last Truffula seed to the child and asked him to plant it. He said something to the effect of "Perhaps if we regrow the forest, the Humming fish, the Swomie swans, and the Lorax will return.". Now that is pretty depressing stuff. I'd rather read or watch "Old Yeller" a few more times at 3 in the morning instead of that.

No time for it. We shall move on!

The Fox and the Grapes

There was yet another great storyteller called "Aesop". He lived about 2,500 years ago in Greece and wrote very simple, understandable stories. I loved his stuff as much as I did those of "Dr. Seuss". He penned one about a Fox and a grape vine. I've never seen a vine greater than six feet tall but that doesn't matter. Let's foray into the story…

A fox was walking through the forest and became hungry. This fox noticed a tall grape vine and it had very juicy fruit dangling from the upper branches. There was also fruit laying at his feet beneath him. The fox tried and tried to jump and reach the higher grapes but always failed and missed. The fox became angry and walked away thinking "Those grapes are probably sour anyway.".

I believe the term "sour grapes" comes from this. For now, we don't care. Just one more thing and we will wrangle it all in.

We have a few nasty parts left to do…

Discourse with an Extreme Bible Thumper

Now we will jump in with a real debate between myself and a hardcore Bible thumper by the water cooler over at an office where I used to work (I will reference this character as **HBT** for brevity). Had our employer overheard this conversation, we both would've been tossed out of the building in less that 15 seconds. Let's move along with it…

Me: "Heya, Phil. Good Morning."

HBT: "Oh, hey Clancy. How are you? How was your weekend?"

Me: "It was alright. The Packers and the Dodgers lost but I got to hang out with some friends. It is all good.".

HBT: "Sorry to hear your teams lost but I am glad you had company."

Me: "Thanks. Did you watch any of those games?"

HBT: "Oh, no. My wife, my child, and I were on a weekend retreat for our church up in the Jemez mountains."

Me: "Excellent. I love it there. My fraternity brothers and I used to party up in that place all the time. We even did some cliff diving. It took us a few beers, *wink-wink*, before we were ready to leap off a 50ft precipice into freezing water though. It is a blast. What'd you guys do?"

HBT: "Well you see, this was really not that kind of party. There was much Bible study and prayer. It was more of a journey for truth and knowledge. My son memorized 3 chapters and can recite them perfectly. He is as smart as a whip, that boy."

Me: "So did he understand it?"

HBT: "Oh yeah. He can recite every line perfectly."

Me: "I mean did he understand all of the metaphors and the allegory?"

HBT: "Ha ha. There is no allegory. This is the written word of God as transcribed by Moses, the prophets, and Jesus with his Apostles. I do not want to offend or judge you but until you accept this, you be will living in darkness. You are welcome to join us any time that you want. Our flock is very welcoming."

Me: "Perhaps I will. There are a few things I would ask your friends if I went though. Do you mind if I bounce them off of you?"

HBT: "You bet! I don't mind. Shoot."

Me: "Well, do you believe that Noah and his family built a huge ship that could hold one pair of every animal on earth?"

HBT: "Yes, and…?

Me: "Well it's just that many animals don't mate in pairs as they are asexual in reproduction. Also, tigers et al need to be fed certain foods. How did Noah and his family keep them off the rabbits? A wooden cage cannot hold a tiger. Where did he store the carrots the rabbits ate and keep the wombats away from those carrots?"

HBT: "It was through God's grace and it was a miracle. We cannot understand all of the mysteries."

Me: "Okay. I will let that drop and you win the point. I am in total agreement with you."

HBT: "Is that it? I have a meeting in 15 minutes."

Me: "Is that meeting about the broken diffusion machine? I have to be there too. I'll just ask two more questions and we'll go to that meeting together. Cool?"

HBT: "Okay."

Me: "Do you believe that Jonah was swallowed by a whale and lived in its stomach for 3 days?"

HBT: "Yes."

Me: "But Phil, the digestive juices inside of a whale are more corrosive than any chemical we use at this company and those that we use can eat through human bone in less than 3 minutes. They could disintegrate an elephant in around 15 minutes."

HBT: "I repeat--It was through God's grace and it was a miracle. We cannot understand all of the mysteries."

Me: "Okay. I will let that drop too. I'm still agreeing with you."

HBT: "I am happy to help and to be of service to those that are lost. God is infinitely forgiving."

Me: "One last question and then we must go to that meeting. Ready?"

HBT: "Yes, brother. I am willing and thankful to answer it."

Me: "Okay, Phil. Do remember the story in the New Testament where a man approached Jesus and asked: 'Rabbi, you tell us to forgive each other. I have already forgiven my brother 7 times. How many more times must I forgive him?'"

HBT: "Yes. It is uplifting. We spoke of it on the retreat this weekend."

Me: "So do you remember the answer?"

HBT: "Yes. Jesus said 'You must forgive him 70 times 7 times.' I am happy I am having this conversation with you this morning. I can see the light descending upon you."

Me: "Thank you. I feel it too. But given the answer, doesn't that mean we need to keep many lists and only forgive each other 490 times?"

Phil gave me a blank stare then walked away to the meeting. He would not walk with me. We never spoke of it again.

That was certainly not a fun way to start a Monday morning but I told you this whole thing would be based on pure logic to ferret out folks on both sides. It was necessary. If it means anything, we were still friends and colleagues later.

Homosexuality

I, personally, am not homosexual so know I am not starting a "movement". I am a Lothario on the best day. That might sound divisive or judgmental but it is not. I said in the beginning that this would be about logic and observation. I am also not trying to give a biased opinion--just facts. This is going to upset some people but it needed to be said. Please stay with it till the end and it will make sense.

The Bible says "Be fruitful and multiply" and "A man may never lay with another man.". Pretty easy too understand, right? Why it left out women is beyond me. Pretty strong words though. I think the pigeons will clear it up. Here we go...

When I was a child, I lived in a very animal-loving family. Many of the kids in my neighborhood had coops. I was sure envious so I asked my parents if I could have one. After their approval, I had me a swell one and I went out to catch them in the streets of Venice. I got 3 of them--two males and a female. My coop was started.

One of the males and the female mated and laid eggs. Those punks would sure be jealous of me now! I had squabs and they grew up. Other pigeons came to see what was going on so I caught and cooped them too. I soon had close to 30 birds in there.

They did not slow down. The cage now had a door cut into the wire so that they could get out. It is hard to tend to that many birds.The problem is that they would go out, bring in more birds, and keep laying eggs.

I began to feed the eggs to the dogs and they had the shiniest coats in our odd neighborhood. There were close to 200 birds then and they were bringing in more. "Be fruitful and multiply" they did. I sure didn't know what to do.

The coop was overcrowded now but those birds did not stop--except for a few. This was odd. Many of the birds were of the same sex. Weird! A 14-year-old can't wrap his head around that. Are they diseased? They did not stop that part either. Not making sense I know. I will get to that and how it pertains to God. Natural population control says me and thank you, pigeons, for the lesson.

Almost done.

Four stories were told and I am happy that I did not say them before the Judge. It would've wrecked my case and landed me as disbarred and in penitentiary. This is supposed to be about God (And it is. That judge has shifty eyes and she does not seem to dig my drift though.).

I do not believe that there was a "Lorax" that hibernated in a tree and was awoken by a cigar-smoking industrialist and his ilk only to defend a non-existent village. I do not believe in "Humming Fish" and "Swomie Swans". I do not believe in foxes that try to eat grapes from impossibly tall vines. I do not believe a pair of every living land animal on earth could live on a large and ancient cruise ship (Noah's family was on that ship and they numbered more than two--not fair to the gophers says me!). I do not believe that people can survive in the acidic conditions of a large mammal's stomach. I do not believe homosexuality is an affront to God. It is why I spoke of pigeons instead of humans. No siree!

I believe in the messages those stories ascribe to though. The Lorax was symbolic of how we should protect this small planet we trod upon as we tend to destroy it.
Simple.

The fox was telling us to not turn our noses up at the things that are freely given to us. Both of those beasts were preaching against greed.
Simple.

The story of Jonah was describing a person suffering a protracted period of depression and darkness--and the solution that was provided.
Simple.

The story of Noah described a cleansing of this planet. This earth has faced many an Ice Age and ice is a form of water. The Grand Canyon in Arizona was carved by a tremendous flow of water that carved the rock apart to form that massive feature-- Don't get me started on fjords. When I was less lazy, I would climb mountains and find seashell artifacts even at their peaks. Noah's tale was all we needed and it is much more fun and interesting to read.

Simple.

The pigeons were in an overpopulated environment so they turned to homosexuality to limit this. All living things need to express some kind of affection and that allowed them to do it without wrecking things further. I do not believe HIV and AIDS are curses from God unless God has it in for birds too. We can stick to War and Abortion for this purpose and continue to persecute our fellow earthlings though. Another method of population control.

Simple.

People cannot wrap their heads around even things that simple. How do we expect them to understand the rise and extinction of Dinosaurs, the Cro-Magnon, the Neanderthal (Well, maybe not the Neanderthal--I believe am the last of them!) and 99% of every other living thing that once inhabited this planet?

This all sounds like a children's book and perhaps it is. I am saying that the Bible is exactly that. Scientists have spent thousands of years trying to counterpoint such stories because they attack the authenticity and facts of the stories but never seem to see how simple and concise the messages are. They are "Parables". Jesus used them, a lot! The HBT use the Bible a lot too to justify persecuting their fellow earthlings. Homosexuals are no different than Heterosexuals--just some do not reproduce. I used the word "Earthlings" because those pigeons fall into the same category. The HBT would walk away again if I said that!

Anyway, I must prepare my case for tomorrow so I will get to that now. I will leave you with a simple question then off I go:

Question

"Do you still believe the Bible is a history book?"

I do--just not the kind you expect.

Chapter 4

Transistor

Before I can enter that courtroom today, I must review a few notes and make sure the Judge has time to read them before I confront her. I will present here some very tedious facts regarding a semiconductor element called the "Transistor". I left out every bit of math I could. We begin. That is a picture every engineer will recognize as a transistor circuit. The thing inside the circle is how a transistor is diagrammed during the design phase of a logic circuit. There are a few weird symbols but I will explain them.

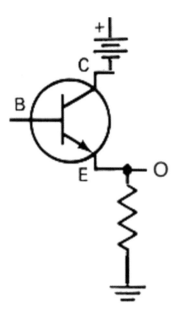

The thing on top that is represented by this...

…is a power source like a battery (It does not need to be a battery. It can be any DC electrical source). The voltage amount may vary depending on the transistor in question. It is kind of important as they will explode in your face if short-circuited. I know this from direct experience and why they had us wear goggles when we had to hook up millions of them.

The thing on the on the bottom represented by this…

…is a Resistor and Ground pair. The resistor limits the number of electrons that may flow through the transistor and also staves off the short-circuit. The Ground is just as advertised. It is tantamount to sticking a bare wire into the dirt.

The B is short for "The Base"(Sometimes called "The Gate"), the C is short for "The Collector", and the E is called "The Emitter". Electrons will flow from E to C when a positive voltage is applied to B. O will be the result. Think of the light switches in your home. No voltage at B means "lights out" or a "0" in computer parlance. A positive voltage applied at B means "lights on" or a "1". The O in the diagram is the light bulb. The difference is that there is no finger to flip these switches.

The transistors operate in tandem and throw each other's light switches at nearly the speed of light. Very complicated math can be solved if they are arrayed correctly--a "Computer". That is for another story though.

This is how a semiconductor operates at a high-level and all we need to know for this part is here. It is far more complex than that but for this, we don't care (It took many thousands of years of science and research before one was created). It is just the concept and facts that are required for that mean ole' Judge and my Atheist friend.

Back in Court

The Bailiff: "All rise".

The Judge enters and seems to be somewhat distracted and annoyed. The first thing she says is "I need to see both of you in my chambers NOW. Bailiff, please escort them.". My Atheist friend and I go in willingly. He seems to be smirking.

We sit down in her office and the Judge slaps down the Transistor report I gave her last night. Here is what happens next:

The Judge:

Counselor, what is this? You are wasting this court's time with your nonsense. It is not amusing. You had better explain to me, right now, the meaning of this. I am running out of warnings and no fruit basket will save you this time.

The Atheist:

I agree, your Honor. I know how a transistor works and this has nothing to do with his case. Perhaps it is my friend here that needs the rehabilitation.

The Judge:

It will most likely go that way or worse if I don't start to hear some answers. Make it quick.

Me:

Yes, your Honor. (Careful not to say any pet-names--that Bailiff is standing right behind me)

I must announce that the document I gave you is completely relevant to this case. The Transistor is a highly complex device and I can tie it into this debate.

The Atheist:

If I may interject, are you stating that it was created by God or whatever fantastic deity you are defending? HA! I know for a fact it was first created at the Bell Labs in the 1940's. It was meant to replace the often-malfunctioning vacuum tube. In fact, the word 'bug' is quite literal. A moth was found inside one of these computers blocking an electric relay. It took them a long time to find that. The word is still in use today. Snuff, God.

Me:

No, I do not think it was created by God--maybe some of the ingredients but that is not the point. You and I can agree that it was created. Correct?

The Atheist:

OF COURSE I DO YOU LITTLE…um…pardon me, your Honor. Yes, we can agree on that. This is preposterous though.

Me:

Did you ever meet any of the people that designed and created it? Do you know their names?

The Atheist:

NO! This was done IN THE 1940's! GET THAT THROUGH YOUR SKULL!

I can get you the names though. I still cannot see where you are going or what this has to do with God.

The Judge:

Agreed. I said to make this quick as it is spiraling into nowhere.

Me:

I'll speed it up, your honor. I just have a few more simple questions that need answering.

The Judge:

Well then proceed. (The Atheist is angrily giggling now. He must've won.)

Me:

My Atheist friend, how would find their names?"".

The Atheist:

In a book. Duh. I can also look it up online. It is well documented. Are you daft?

Me:

Good answer. "A Book" and "online". It is perfect for my case.

The transistors "created" today are so small that it takes a fairly powerful microscope to see them. The fact that you said "online" goes to show that you have had sessions with a computer. I see you have a cell phone too. Love the fancy little holster you got for it!

The Judge:

Get on with it.

Me:

Well, that computer you use and that cell phone are built with millions of those invisible transistors. They communicate in wavelengths of light we cannot see and they use electricity to operate. You cannot see electricity either. They operate in near-perfect harmony to achieve their function.

The Atheist:

And? Are you telling me that a God is responsible? Your Honor, please?

The Judge:

This is your last chance, counselor.

Me:

Sorry. Almost done. No, I am not saying that it is the work of God. We all agree that these highly complex systems were created though. My Atheist friend said the word "created" first. I learned about creation in a book too--a very old and different book.

The Atheist:

So what?

Me:

Well, if we are in agreement on that, then why do we disagree on the point that the universe, far more complex and harmonious than any system of transistors, was created? That is a big part of my case, your Honor, and the reason I sent you and my esteemed colleague that document last night.

The Judge:

I see now. I will allow it.

The Atheist:

BRAAGGGGHHH!!!

Luckily, that Bailiff was there. My Atheist friend seems to need a few more days at the "happy" place. God bless him. Court was adjourned for the day.

Albert Einstein made a historic quote that said "*Energy cannot be created or destroyed--only transformed*". My Atheist friend will refute that and say that our bodies will only become food for worms. There is no "Soul". There are about 1000 ways to counter that and he better be sure he is right! We will get to that in the 6th chapter of this wacky book.

In the next chapter though, we will press on with a similar conundrum. It will involve some science, arts, and crafts! YAY!

Chapter 5

Harpoon

We have a few days away from the courtroom now. The Judge sent ME a fruit basket and a note of apology. The ice is finally melting. I also re-gifted the pear and had it sent on to the institution. I already know that they will not let the Atheist have it--Something about how it will interfere with the sedatives he is receiving. I am no doctor so I do not know. It was the act that was important and I hope he feels better.

In the interim, let us play a fun game. This is a game scientists and theologians have been traversing for many thousands of years. The rules are simple. On the next page, you will see a picture. All you have too do is uncover every pattern and rule you can observe.

I will leave a few examples and 3 blank pages should you want to jot down your own answers. When you are done playing you may choose to read along to find the true solution. It will have some pretty impossible questions whether you be scientist or not. Let's get it going!

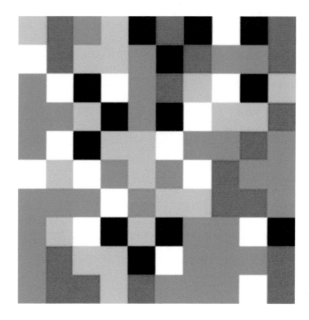

It appears to be a set of squares with random colors assigned. If you look at it, I'm sure you will begin to see patterns emerge. I will give you a few of the obvious and easy ones. You will have 3 pages to figure out what it is and why it looks like this.

Examples

- If there are two green squares that are disjoint on any row, it is guaranteed that there will be two adjacent green squares in the next row.

- Counting from left to right starting at the number 1, the black squares are at odd numbers in that row. In row 1, the black squares are at 5, 7, and 9 columnar. On the even numbered rows, they are in even numbered columns.

- If there are two red squares adjacent vertically or horizontally, you can look diagonally in every direction and you will find another red square in the grid to match each of those on the adjacent squares.

You are already bored I can tell. Like I said, you have 3 pages to scribble on after this though. Give it a crack and keep your aspirin handy. I know you can find the answer!

Whew! There are a lot of patterns and math for you to derive to help you describe that image. Those 3 pages were nowhere near enough.

I also sent this over to the institute as my atheist friend likes to solve puzzles. The Doc says it is therapeutic for him. Anything to help. All scientists love to unravel and put order to problems so I thought this would be a good place to start. God bless him (Yeah, I said it again).

I hope you found the secret answer. This is certainly not an easy one (or perhaps it is easier than you think!). We will see how much the pattern-matching and math helped us to get closer to our truth. Please take as much time as you need to think about it. As was said, those scientists and theologians were given centuries to figure out such things. We are just talking about a 10 by 10 grid of colored blocks though. It should not take long to reason it out.

Turn the page when you are ready and the answer will be unveiled.

Laverne and Shirley

Before we get to the answer proper, I need to get this out of the way:

When I was a child in the 1970's, there was a very funny television show called "Laverne and Shirley". It was about two young girls living together in Wisconsin trying to make ends meet by working in a brewery. It was set in the late 1950's or early 1960's. This show was a spin-off of another classic comedy called "Happy Days".

Laverne and Shirley had a few interesting friends but my favorites were "Carmine"(The usual hero) and "Lenny and Squiggy" (The goof-ball antagonists with the greased-hair look).

There was an episode where Squiggy did something stupid and Carmine was coming over to give him a beating. Squiggy probably ate a piece of Laverne's birthday cake when nobody was around. Carmine never did harm to Squiggy as Squiggy apologized in a heart-felt moment and Laverne forgave him. There were many other similar episodes but that was the general gist. Fun show.

Since this was network television, they had periodic commercial breaks-- All of the "Hey Mikey, he likes it!" '70s stuff. I also remember "Calgon" and "Clorox" commercials as well. The television reception then was all transmitted by airwaves. It was less than a perfect way to achieve this but, besides a little static, it did its job.

This sure sounds like it is meandering into nonsense again, huh? Although I made amends with that Judge, I know she would be looking at me sideways about now.

Yes, to the point!

You can refer to the math you penned before but I can guarantee you this-- nobody in a thousand centuries would've got the answer right. From the Greeks to Mr. Hawking you could end up with some interesting answers though.

The weird grid of squares you looked at before, and there were only 100 of them, was a 10 by 10 sample of a random location of output from a television screen. This was from a single frame of a "Clorox" commercial during a break from the "Laverne and Shirley" show. The red, green, and blue squares were natural emissions from a precisely calibrated Cathode-ray tube that allowed animation to register in your brain. The black and white squares are "static" and they can be produced in many forms from sunspots to "Dad didn't put up the antenna right.".

"So what?" Well, we'll say "what" now…

From your math and descriptions, you never would've figured out that this was a small snapshot from a commercial during an American television show. It would not have explained the circuitry and operation of a color television. It would not tell you what show was airing at that moment. It will not advertise "Clorox" to you in the slightest way.

It will not tell you why Carmine wanted to put that beat-down on Squiggy or why Laverne was angry about his transgression. It will not tell you about the folks that created it or what technology they used. It will not tell you about the people that met and decided to "green light" the filming of this show. It will not tell you the names of their family members or their intentions. I could make this into a list of a million more lines. It is not required. You are starting to get the point.

Scientists examine grids like this a googolplex more complicated in the universe but their pages are no better than the three I allowed you. This includes the Englishman that strolls around in a talking wheelchair and proclaims "No God was required to create the Universe.".

Putting a hole in that atheist arrogance is one of the purposes here. Remember, I am not trying to prove there IS a God. I am only trying to show reasonable doubt to my sick Atheist friend that he better be careful about what cliff he chooses to jump from. Science is just a method of observation and I enjoy it too but it is certainly not a way to disprove God.

Anyway, I am awaiting a call from the court to see if we convene tomorrow. The people at the clinic also say that the Atheist is still not out of the woods yet. Maybe this means I can sleep-in come morning.

Question

"From the data you were presented, can you tell me the name of the dog that Squiggy's stepmother had when she was a child?"

While we wait and you refer to your math to answer that, I say we jump into a few fun and personal ghost stories. They involve members of my family and me. After each one, I will try to dismiss them logically as the Acting Atheist. It is only fair considering the crazy one-sided nature of this book. Next chapter…

Chapter 6

Ghost Stories

A few calls later and it seems that court is cancelled for today. I was able to sleep till 10am. Fine by me. That "All rise" thing nearly aggravated an old football injury. I know I whine too much. That silly lifetime pain I earned tackling a guy twice my size cannot rival the treatments my Atheist colleague will undergo this morning. That will sure be something. The Judge has a hair appointment today anyway. I know these things as I have my spies too.

Let us get into the real fun stuff while we wait in the wings. I have a few ghost stories for you which I, the Acting Atheist counselor, will debunk to the best of my ability. You are welcome to do the same. I have left out about 40 or more such stories but I think these will get the point across.

Disclaimer:

They are not scary and they certainly will not give you nightmares. "Doubting Thomas" will be right there by your side for the whole thing.

The Parrot

My wife and I used to live in a nice apartment very close to the Pacific ocean. Everyone that also lived in this building were very close to each other. Each morning or night there seemed to be an event that gathered us all and always some intrigue brewing. We were always welcomed to each other's homes. It was very much "Melrose Place" and sublime--in fact, I think the writers of that show could've taken a few stories from there. Many doozies!

On one day, a woman moved in. She was a little bit older than us and probably was an ex-actress in the "adult film" industry or something along those lines. She was of the tipplin' way and when I say "Tipplin'", I literally mean a gallon of wine each night and day. It is her life and she may do as she pleases. Let us just refer to her as "E".

She brought with her a bird. It was a smaller parrot and called a "Cockatiel". It was very friendly and she seemed to take good care of it so nobody said a word when it screeched all day and night.

E did not hurt or mistreat this bird. It was a very happy. She was asleep 16 hours of the day anyway until the wine wore off. Somebody else was paying the bills.

One afternoon in summertime, a Baseball game was being aired on television. It might've been the Dodgers and the Giants. I made myself a big greasy sandwich, cracked a beer, and found a comfy spot to relax and shout "YOU CAN'T LAY DOWN A BUNT? YOU IDIOT!". My wife did not like that kind of thing so she went out shopping. Perfect and I was getting a glimpse of heaven. Oh, and I was wearing cargo pants, a tee-shirt, flip-flops, and a ball-cap that should've been retired 5 years before. Bliss for a guy and understand that, ladies!

At about the 4th inning I was going to shout out something else when there was a knock on the door. I went to answer and there were three or four ladies from the building. I said "Hey, is the volume too loud? I can turn it down. I think I hurt my eardrums at a concert awhile ago. I am very sorry.".

The ladies said "No. This is different. E came home from a bender and left her door open before she passed out.". I said "Is she okay? What do you need me to do? Not sure I can help.". They said "She is asleep now. Her bird escaped and it is loose in the neighborhood though. We know you can find it.".

I protested and said "If that bird got out, it is gone-- as I looked over my shoulder at the TV and screamed "YOU JUST SWUNG AT A PITCH IN THE DIRT 4 FEET FROM THE PLATE! WHO HIRES THESE CLOWNS!".

End of an inning so I went out with them and I was grumbling. I got zapped by something from somewhere else and I have no idea what it was. There are more than 500 bushes on the loud street I live on but I went immediately to one that was around two and a half blocks away. I said "The bird is in there". I unraveled the branches and there was the wayward cockatiel.

I grabbed it very quickly and, yes, it did bite--very well, I might add. Those ladies looked at me in odd ways after that. It was not me. It was something else. I got to see the last 4 innings of the game before the wife came home with $300 worth of shoes. Weird but who cares? What does that have to do with God?

Acting Atheist Rebuttal:

The bird must have had atrophied wings from being in a cage for so long so it could not have flown very far. It does not matter how loud that part of your town can become as humans have great hearing--not as good as a dog's but good enough for this. Baseball is boring and it puts me to sleep too. The ear does not sleep ever even though your brain sounded pretty close to that.

You heard this hungry bird squawking over all of that din and you tracked it to that bush. It was not a "lucky first try guided by a spirit". The ear is very complex and well documented in its abilities. Please acknowledge that and please quit the crazy talk. Move along.

Cody

My dearly-departed mother was a great animal lover; She continuously had around 30 of them near her. There was always a "chosen one" and in the parlance of the Wiccan, a "Familiar". There were many favorites from her childhood spanning from one-legged chickens, cats, dogs, and many other varieties--even a crow, for crying out loud. The "chosen one" was fiercely loyal and guarded her in an extremely frightening way. The last she had chosen, before her soul decided to leave its earthly cage, was a dog named "Cody". This is that dog's odd story.

Cody was a small dog called a "Sheltie". They look like a regular Collie but about 5 times smaller--show dogs I think but this is not about that. This dog was very quiet and deliberate. I rarely saw it more than five feet from my mother. If one got too close to her, Cody would give this "stare". The eyes were black as night and they never left your gaze. Even saying "I love you, mom, but I gotta go." then trying to give her a hug and kiss seemed dangerous--the quiet stare throughout. Cody never blinked once.

So what?. We all know that dogs are loyal. It sounds like this one had a personality. My sister's dog 'Peter' used to…

Please wait, my Acting Atheist friend. You will get your chance in a moment…

I live in a suburban neighborhood and have been for many years. There are raccoons, possums, squirrels, and, once in a while, the occasional rat. It comes with the territory and I have done battle with all of them. They will come inside if I am lazy but I can get them out in less than 10 minutes. It is all good. That being said, there is absolutely no access to the attic. They never have had it and hopefully they never will.

I must repeat, SO WHAT? We get them around here too. 'Peter' used to…

Acting Atheist colleague, please let me finish. Cody is still on the stage. We will get to the end of this story now and you can have your rebuttal.

At night, I like to finish it up with a movie starting around 10pm and in bed by midnight with a book read. I was doing exactly that once of many times and the two cats that choose to live with me were variously draped on my lap. I don't remember the movie but I'm sure it was good as I'm a snob when it comes to that stuff. Everything was just right until…

It was 11:45pm and the movie was running credits--sleepy now. Then suddenly I hear three loud bangs and a snarl in the attic right above me. The cats jumped up onto the table and were staring right at the spot where it happened. I said "Ah #$^%!". They got up there somehow. I went and got the broom. I hit that spot hoping to hear something running away. In this way I could detect the entry point. Nothing. I went out with a flashlight and checked every grate on the building but they were all solid. Maybe they were tunneling in?

I went back and was undressing for bed. In the closet, "BANG BANG BANG" and the snarl. The cats were at my feet and staring straight up again. I got the broom and hit that spot too. There was still no sound of feet running for an exit. I know what those critters sound like when they are scurrying. Oh boy, do I ever.

GRRR…I went out with the flashlight again. Nothing.

I was dressed for bed now so I hopped in and turned out the lights. After that, I had no energy to read books. Right above my head in the bedroom "BANG BANG BANG" and the snarl. The cats were there for the third time staring straight up. This time I did not go for the broom or the flashlight. Instead, I thought a quiet whisper in the dark: "You must leave now. You are not welcome here.". I stayed awake until 12:45am before I finally snored. It did not happen again. It has never happened in the 4 years I've lived here before and it has not occurred again.

I called my sister in the morning and told her all of that up there. She was not surprised or concerned at all. She said something like "Clancy John, Cody passed away last night. You should not have said your little prayer. That was a 'goodbye'.".

My sisters came in the morning and we triple checked the place including the incredulous neighbor's home. No tunnels. There is no access to that attic for man or beast. Your floor, Acting Atheist.

Acting Atheist Rebuttal:

You live in a part of the country that is prone to earthquakes, yes? Those things bother me too so feel no shame. The tremors created are often directed into localized spaces. Many people do not feel them even though their neighbors might. In this case, a few spots in your home were affected. The "banging" and "snarling" are usual sounds an earthquake makes. We have different technical terms for them though.

Animals can sense these things and they act differently when this occurs. It has been noticed with gerbils in the laboratory. We are not sure how but we believe these animal abilities exist.

You experienced three aftershocks and you, in your sleeping state, did not understand so you resorted to brooms, flashlights, and prayer. It is all good. We know why the events you described happen. It is called "Geology" and another subject you should read up on before you bring these things before us. Okay? I thought we were here to talk about "God".

Okay, my Atheist brother. I was born here and lived most of my life here. I've been through many earthquakes--to the point that they don't even bother me anymore. What happened on that evening was no earthquake. You are correct though. I only took 2 semesters of Geology in college so I understand that I am no expert. Perhaps there is a rare phenomena that they did not teach us about and I got to see it for the first time. I will let that drop and you win that one. I will go on to the next thing now.

Guardian Angel

I have another sister. She has experienced things that most haven't. I'll tell you a small one then move along to that "Guardian Angel" thing.

She was driving down a street one day as a passenger and chatting with the driver. It was a very nice day and I bet their conversation was very interesting. She has a way of talking that makes people think before they respond. I am sure there was much debate and laughter.

On this day, a woman stepped out from between two parked cars and was summarily run over by that friend of hers. This woman bounced off the hood and windshield. She was pronounced DOA and my sister was in near-panic. When my sister calmed down, she certainly had a lot to think about and surely traumatized. She is definitely very difficult to drive with now so at full attention you must be. Don't mess up behind the wheel with her around.

Acting Atheist:

What does that have to with Guardian Angels or God? You are meandering again and it is getting old. Say what you are trying to say and come on...

Atheist fella, I am getting to that. You will get your chance, smart-guy.

One evening, my sister was going out to a prayer meeting or a party or something. She tried a jay walk across a busy avenue. Long Beach, CA is not a good place to attempt this feat. For some reason, she felt two hands on her shoulders and they pushed her onto the sidewalk behind her. A split second later, a bus came rushing past that would've killed her immediately. There is no way a human could've done this else they would've been the lasagna we would scrape out of the street in the morning. Something shoved her. I'm sorry that did not happen with the nice lady my sister's friend ran over. That lady was older and perhaps it was her time. I believe my sister but I will give the Atheist challenger a shot.

Acting Atheist Rebuttal:

You still really do not understand science. Large objects moving at high velocity displace the medium in which they are traversing. You can see this when a large ship passes by and there is the wake left behind. This also happens with airplanes and yes, with buses. This is a force perpendicular to the direction in which the object is moving. Your sister felt this force and perceived it to be two hands pushing her away because of her Catholic upbringing--nothing more.

As for the old woman that was tragically killed, the car that struck her only needed to be moving at 30 miles-per-hour. Also, cars are much more streamlined than buses so they do not displace enough air to move the weight of a human body. I hope you are understanding this better now. Do you have any other questions that you would like explained, sir?

Yes. I have seen large boats and buses in action and I agree a very strong wake is produced in their passage that can push things away. I have never seen a wake precede their arrival. Explain that?

Acting Atheist:

Well uh, you see, um…

Me:

Exactly what I thought you might say. Your "certainty" seems to have abandoned you. I will not hammer on that anymore. I believe I am coming closer to the answer.

Just a one more story to hear and I will give you a chance to refute them using *your* logic.

The Bug

I know we've already been through a boring section that included a shape-shifting moth that submitted to the Theory of Evolution in the span of 30 years. I also mentioned that I tinker in computer-related work so let us begin with a "bug". Here is a simple example of an actual one that computers must defeat:

```
void FillArray(int* array, int numElements,int value)
{
        for (int iter=0; iter<numElements; ++iter)
            array[iter] = value;
}
```

I can see about 10 ways that this thing can go wrong (just like your argument) and crash the entire program. Functions like that are interspersed in hundreds of thousands of lines of code and sometimes it is hard to catch them misbehaving. A boss breathing down your neck saying "Why is it that every time I start this program, the screen turns blue and I have to unplug the machine? It always does the same thing.". We don't care about that, boss. I will speak of a far different "bug". This story is quite different.

I put that weird piece of computer code up there for a reason. I work with that stuff constantly and those machines do not feel emotions. All they understand is "Black or White", "Yes or No", or "1 or 0". They are very harsh and they do not care if you are having a bad day. Insects have more emotions than a microprocessor and those insects certainly do not mess around. No "hugs" there.

Acting Atheist:

I agree with you. What does this have to do with God? I received my degree in…

Me:

Let me continue, my Atheist buddy.

A while ago, I mentioned my mother's soul decided to leave the body in which it was housed. The family was pretty wrecked and we all were for quite a time. I wanted to walk the plank too but I know my mother would never allow that. She was a computer programmer herself and we had many, many, hours debating algorithms. Interesting times!

I was in front of one of those machines one morning--insectile mind in full effect. I was killing it. I got this weird pang over my left shoulder that I could not shake-- I felt my mother near so I called my sister to commiserate. (She and I can talk for many hours but this one was relatively quick -- 45 mins?).

After this time about talking about our mother, she said in a very bored way: "Clancy, expect a visitor today. It will come in the form of animal.". I said "Yeah, yeah. Gotta go. I love you and I will call you soon. Thanks for talking to me.". She said "I love you too. Goodbye.".

"Hocus Pocus" and whatever, sis. She had obviously lost it. I was sad for her as she was hurting too after that talk. It meant more work for me as well; I just blew 45 minutes of my life talking nonsense and upsetting my sister.

I returned to the insect world and trying to figure out the fastest way to shuffle different screens programmatically on an iPhone. I was back in "The Zone", baby--code head. I was?

Less than 5 minutes after ending the phone call, I felt a tickle on my neck. I grabbed at it immediately and there in my hand was a ladybug. A ladybug? That just does not happen here ever. I was not going to hurt it so I rushed out to put it in the garden. When I went out the back door, there in a tree by the porch were two squirrels, two bluebirds, a big white cat, and a flock of sparrows. Now these animals do come around in phases but if anyone of them ever met, they would try to kill each other. In this case, they just stared at me. I put the beetle in the bushes and went in to get them some food. When I came out with it, they were all gone.

I called my sister as fast as possible to tell her. She was too bored to pick up the phone but when I did get ahold of her, she just gave me the "I told ya so.". My sister lives 17 miles away so there was no way for her to have set that stage especially with Los Angeles traffic. She just "knew" and again was bored. Explain that, my dear Atheist amigo.

Acting Atheist Rebuttal:

Listen to *me* for once. You admitted that these animals come around regularly, yes? Were you talking on a cell phone? The cell phone communicates with waves of light that you cannot hear or see. You already beat that into the ground before. Animals have different perceptions. They can hear and see things that we cannot. They heard this conversation you were having and appeared together expecting to be fed.

The reason they did not all try to kill each other is probably because they had determined a pecking order. This is common in animals. Also, the ladybug you mentioned is not a rare species. I've seen them many times before--not in an office so I'll give you that.

Anyway, what you witnessed was a coincidence. The fact that you were talking to your "spiritual" sister is silly. Animals do not speak your language but they can sense the vibrations your phone emits and even your tone. This has nothing to do with God or anything. I'm not sure why I'm even wasting my time with this vitriol. What do you say to that?

Me:

Well, Acting Atheist, I use that phone quite a lot and have done so for many years. Never before did anything close to that happen. A "coincidence" you say? I will grant you that back. I really do not want to fight with you. I do wish I had that coincidence luck in Las Vegas though. You betcha.

I agree with you that what was said up there seemed to have nothing to do with God. We are not in the courtroom so the Judge and your counterpart in the hospital cannot get me. I have another query for you.

Question

"Is there a reasonable doubt of your assertion that there is no spiritual plane beyond that which we can perceive?"

Your answer is all I need for this case but I already know what it is. Your "Well uh, you see, um…" is good enough for a winner.

Now you see where we did defend God. Closing statements in the morning so 9am for me. This should be good. They brought in a another atheist to nail and win it all for the prosecution. FUN!

Chapter 7

Summation

We are back in court today for the last time regarding this case. The Judge is here and on time. I saw her on the steps. Her hair looks great (I think she had it frosted) and she is smiling. I shot her a wink and she winked back. It is looking good so far!

The original atheist in this hearing was not able to attend. The doctors say it takes many months and sometimes never to recover after they sever the prefrontal lobe in such a complex medical procedure. The reports say he is calmer now and speaks much about rabbits, butterflies, and funny shapes he sees in the clouds when they bring him out on his walks. I don't think any more pears or puzzles will help him now though. Ah, we tried…

The Acting Atheist is caged in a single brain cell in the back of my mind. I only have three of them left now so know that I am being generous. When that one erodes from the "Fatal Glass of Beer", I, he, and his cage will go back to the enormous energy pool that is the Universe. I can't say I'll miss him as he almost got me. He is a smart fella that one--great points.

They assigned one last one for our day of summations as it is proper etiquette. He is a bit younger and a little jumpy but he seems to have something to get off his chest. A few more mutual winks and batted eyes at the Judge as she prepares her entry so I decide to let him go first--more time for the flirting!

The Bailiff: "All rise."

The judge now enters and announces the agenda for the day. We each will get a turn to give closing statements and an urging to conserve time.

She explains to me and to the jury that this new counselor for the prosecution is to be known as "Atheist 2.5". That makes sense to me considering the state of the original. We should of had it at "Atheist 3" but I choose not to go into any more detail concerning the plight of "Atheist 1". It is one of those people we should show compassion for but not really talk about very much. The way it is.

Anyway, everyone can sit now and we can finish…

The Judge:

Mr. Prosecutor, you may now give your closing statements. This court apologizes for your late summoning and will allow you all the time you need to speak. Please proceed.

Atheist 2.5:

Thank you, your Honor, and please notice how I said that properly unlike that miscreant over there. Ladies and Gentlemen of the Jury, I will begin.

I have read all of the notes from these proceedings and I still do not see what he is saying. My colleague tried to throw in the Bible, some mathematics, and a few insects up as a smoke screen. He also tried to show you an odd picture and then relate that to an old television program--"Laverne and Shirley"!?! He even talked about ghosts and children's books. Are you kidding me?

My friend here is trying to fool you. He has tried to mix up many things to put them over your head that you might find a verdict in his favor.

He is a charlatan of the first water and would be better off selling snake oil in some desert and soon-to-be ghost town. He can tell his stories to them--I hope they do not conduct a hanging. That is what they did in the "old days" with folk like him.

Everything this person has told you was a lie or otherwise fictitious. He has nothing and I know you are all smart enough to see that. It is the main reason my now-infirm associate chose you to be on this jury. I would've done the same. I can see it in your eyes.

I will finish up by saying one last thing. He has not proven a "God" or anything else. He sure took up much of your time though. The likes of us are paid by the hour--myself included. You on the jury were taken away from your homes and your families when we were getting wages bickering over unwarranted and ridiculous fallacies. That "man" over there dragged it out as long as he could. We should not be here right now. I would rather be on the beach with my friends and family. Life is too short for this kind of thing.

Think carefully with him and his tactics as the next thing he will probably show you is a card trick. Unbelievable. Vote wisely and I will get you back to your children by the afternoon. I will let that serpent speak now so please remember my words. I thank you and apologize for your time wasted.

Your Honor and the jury,

The Prosecution rests.

The Judge:

Thank you, Counselor. Your thoughtfulness for the well-being of the families of the jury is duly noted. You were brought in on short notice but you said the things that many of us were also thinking. I have noted that too. I would like to remind you and the members of the jury that this case is NOT a proof for a God. Please keep that in mind.

In the future, I would stay away from the name-calling. It might just be what your opponent wants from you. It sure seems like it. He will give his summation now so, Counselor for the Defense, present when you are ready.

Me:

"Two of Hearts".

The Judge:

What? What kind of summation is that?

Atheist 2.5:

Yes. What? It is the drivel he has been spewing throughout this entire trial. I'm telling you he is a liar. What does that have to do with anything? THAT IS WHAT HE DOES!

Me:

My friend asked for a card trick so this is a gift for him. I thought he and everyone here might like it. It's a good one.

I would ask him to reach into the left front pocket of his shirt and show the contents of it to you and to the jury.

Atheist 2.5:

Okay. It is a playing card with the suit and number that match. This only proves that he is a burglar of sorts. How he knew what shirt I would be wearing shows that he is involved in a conspiracy down at the dry-cleaner or something worse. This is criminal.

He is evading his proof of this "God" with stupid parlor tricks. He is only trying to confuse you and keep you here longer. It is making us all weary.

The Judge: (*bang bang goes the gavel*)

That is enough, Mr. Prosecutor. You were allowed to give your summation. If there is some proof of foul-play then we can meet again for a separate trial. I would like to know how he did that though. Would you please explain this result, Counselor for the Defense?

Atheist 2.5:

Yes, your Honor. I have the oddest feeling we will all be meeting together again soon. That guy will pay for this. We are no where close to being complete.

Me:

    Your Honor, I am really not required to divulge how such tricks are performed. My opponent's rendition would be a great explanation though-- I must write that down. It would be a good way to do it. Anyway, the Magician's code, chapter 42, subsection 7, and codicil 3 says…"

The Judge:

    Yes, I know that code--"A Magician is not required to reveal the methods or circumstances of any trick or illusion.". I was just curious. Finish your summation without any more funny stuff please. Continue.

Me:

    Thank you, your Honor, and ladies and gentlemen of the jury for hearing me out. I will try to make this quick. The beach awaits!

The "card trick" was in defense of our "Laverne and Shirley" argument. It certainly has my atheist friend thinking now. At least he is wondering "how". This will make more sense in a few moments. It also shows that no magic was involved. The trick was all done with a well thought-out plan and by design. An "intelligence" was required--in his opinion, the universe was not.

He also said that he wanted to get you all out of here quickly and that we were wasting time. Because of that simple trick, he formed his own conclusion (it was incorrect), passed judgement, and has us all slotted for a new trial date. Perhaps he is the one playing with false concepts here.

I do not mind the names that he called me. He was out of material and was limping on a last leg trying to cover for his fallen predecessors in this trial. I would do the same just as a wounded dog would do if cornered and defending her pups. I hold no animosity towards him but please realize that what he said has nothing to do with this case. It was an emotional outburst and nothing more. I have had those too as I am sure you all have as well.

I am glad he read all of the notes from that other atheist fella down at the clinic trying to tame and and assign names to the Carp in the "Serenity Pool" over there. It is honorable.

He was correct about most everything. I did bring up the Bible, math and science, bugs, kiddie stories, "Laverne and Shirley", and the ghost tales. The kiddie stories were for us all including those in the Bible. They were simplified to make it easier for us to understand.

We all seem unable to grasp this. Aesop, Dr. Seuss, and Professor Einstein are all nodding their heads in approval now I can tell you. They were trying to make understanding simpler for us too.

My opponent has made a huge blunder though. I was never asked to prove that there IS a God. My job was to put doubt into the minds of those that propose with 100% certainty that there is NO God. I have never made any claim otherwise. That is the part of the case-notes he forgot to read. He only had a day to review all of this information though--a "day". Hmmm. I don't want to get started on that again.

It is perfect weather outside at the moment so I will try to finish this up now.

At first, I did some boring math but it was required. I tried to avoid this but without it, nothing would have made sense. Grasping a few simple concepts, and I made them as simple as possible, were important. Some people just think a "Black Hole" is a theoretical buzz-word that eggheads use. I had to clear that up with a short science lesson.

I described some discrepancies in the concept of time. The Mayans had a different calendar than that of the Romans and on and on it goes. Oh, and I am only talking about earth time. It makes no difference to a Cesium atom's frequency and the harmony that it measurably exhibits. I believe God created this atom but then again, perhaps I am a lying confidence-man as my friends contend. We set our world clocks with these atoms though. Let them deny that.

Earthlings on this small blue marble cannot agree but some presume to speak for the trillions and trillions of other planets that are flying around out there. Is it creating doubt yet?

I moved on to show that differences in language and the concept of time were the only qualms between Creation and Evolution. I explained how day and night were metaphors for "beginning" and "ending"-- Pretty easy to grasp I think.

We talked briefly about the hard-and-fast proven Thermodynamic law of Entropy and how energetic systems tend towards chaos. We also talked a little bit about the details of Evolution and Genetics--both of which mysteriously violate this law. Again, an "Intelligence" seems to have been required. The car you drove to this courtroom this morning did not create itself. I do not hear any objection to that statement--not even crickets.

We travelled on to show how fantastically written stories contain hidden meanings to make them simpler for us to understand. The Bible is brimming over with such stories yet some take them literally. It is why I called the Bible a "children's book". It is for teaching and not a true history record. With that I agree.

I am a Christian and I believe in God but having that conversation with the Bible thumper, and it is a true story, ripped my heart apart. We are all seeking the truth and not another reason to polarize ourselves from each other. Intentionally hurting that person was probably wrong. That conversation was required by logic though and perhaps gave him a thing or two to think about.

The argument over the "Transistor" was to show that there are things that we cannot see or understand. We all have faith that they were created and that we can read about these creators in books. We have never met any of them. We only have faith and evidence. I also have faith and evidence for a creation event. When I go out and look at the stars at night, the tree outside, and that fella on the corner I am always trying to knock out, I believe they were created. I must repeat, that idiot card trick was not "magic". It also was created beyond the means of my friend over there to comprehend. That "Transistor" analogy was meant to show further doubt of the atheist ultimatum.

The pretty picture I showed you and asked you to explain was to prove that a snapshot is not enough data to define "everything". It was a harpoon shot to the whole atheist bit of nonsense certainty. I probably could've just presented that question only and we would all be at the beach right now. For this, I apologize. I just wanted to make sure you understood why.

The odd ghost stories were not meant to scare you. They are interesting and hard to explain. As I said, those were just a few of many but I am trying to make things clear as fast as I can. I gave the Atheist a chance but no solid arguments came from him. I will really finish now. The last thing I want to say to you is...

WAIT!

HEY LOOK! 2.5 has torn the card to pieces and he is grumbling in some other language. He is trying to pull his hair out too, your Honor! Wait, that is not grumbling. It sounds like a Gershwin tune. One of my favorites!

The Judge:

Bailiff, please assist him. Take him for a drink or a coffee. I also have some light sedatives in my chambers. Just give him one. Counselor, please wrap this up.

Me:

Yes, your Honor. I apologize to everyone here. He has a stressful job trying to attack a case he is sure to lose. The pressure must be terrible but they have established ways to help him. Why would they help him if there was no God to answer to at the end of our lives? Why not just put a bullet in his head and get him out of his misery right now?

He just wanted to go to the beach with his family. I have faith that he will be okay but I still feel terrible. God bless him (Oops! I said that again) . I will put my hand on the Bible here in this courtroom and swear by it.

The Judge:

One moment, Counselor...

(Scrabbling and singing from the back of the courtroom as the Bailiff attempts to remove Atheist 2.5. He forgot his toupee that he had ripped off! I won't sue him for the card he destroyed. It was from his own deck!)

You may continue now. This circus has gone on long enough.

Me:

I agree it was a circus and of the three-ring variety, your Honor--one for each of the atheists that shadowed this room. We will kindly show this tolerant jury the way to the egress now.

Members of the jury, if you believe that I was here trying to *prove* a God, then your decision will be clear. Everything I presented to you would be absolute nonsense as those other fellas continued to state. It may indeed place my neck in a noose as well should I get too close to a church.

If that is what you decide, then your deliberation must agree with the arguments of the first atheist who is now playing checkers against a laundry basket full of cabbage and Atheist 2.5 that was just dragged out of here singing "Someone to Watch Over Me" in Japanese.

Every moral fiber in your body must agree with them. You will have no choice.

BUT!

If you believe the true purpose of this trial is to put even a small modicum of doubt into the atheist ethos of "We know everything that we need to know. There is absolutely NO God or creator. It is a complex bit of mathematics and physics that can define it all.", then all of the evidence I provided to you will make more logical sense. Remember, we only need 1% doubt.

Before you go deliberate, please note that unlike the rotating door circus at the other side of the room that I believe the judge was alluding to, I was here for you from the "beginning" and now the "ending".

QED

Before you go to decide, I have but one last question
and it is the most important of them of all.

**"*Can you prove to me or anyone, with 100%*
undeniable certainty, that NO God exists?"**

Thank you for your time and enjoy the beach.

The Defense rests

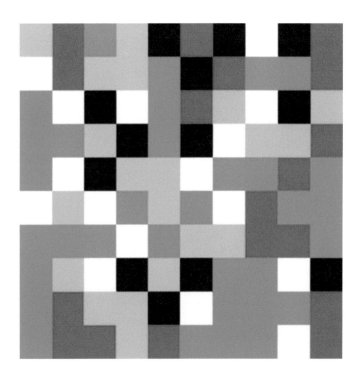

Epilogue

Thank you for reading all of this. I said on the cover that this was a "math book". It truly was but except for the odd equations here and there, I avoided the conventional notation for the reason that you would've thrown it in the trash bin already(and perhaps still will).

I know it became very repetitive and redundant at times. This was done on purpose just as it is done at every level of learning. Sometimes people need to hear things more than once before they "see". The Space Shuttle has several redundant computers on-board so that if one fails, the next one will pick up where the damaged device left off. "Don't put all of your eggs in one basket.". This is common sense and logical.

I said in the prologue that I was a scientist and that I also believed in God. To clarify, I was raised in the Catholic Church and if you want to talk about repetition, try performing any Catholic rite from a Rosary to an Exorcism. I suppose I let some of that bleed over into this document but I feel like I sheltered you from most of that and only used logic. I told you this was not going to be a religious text. See? I am repeating myself again!

I hate protracted epilogues more than long prologues so I will set you free from all of this. If you are interested, you may turn the page and learn about the characters included here though...

The Judge

She is a very austere woman. Her goal is to enforce the law and she is very good at it. It took her many years to reach this point with much laborious study. My wisecracks toward her and her admonitions were all the proof I needed there.

Note that the "laws" she protects are the product of a small species of many living on a tiny planet in the faraway arm of a galaxy called the "Milky Way". It is just one of many trillions of galaxies out there. Yeah--I said "Trillion" again. Also know that the name "Milky Way" was assigned by that same small percentage of the creatures living on this remote planet. That is her life.

God bless her.

The Bailiff

He does not speak very much. I've only heard him say about 12 different words. He is faithful to the court and does exactly what he is told to do. He follows the law to the letter and is a perfect counterpart for the judge. His presence in this book is exactly the reason I included the debate with the Hardcore Bible Thumper. He seems a good man and is never late for anything as his understanding for time on earth is uncanny. He is very focused. It is the akin to why blinders are put on horses, why the falconer puts a hood on his bird, or why baseball players wear caps--pure focus. His personal belief in God does not matter. He extends the Bible and gets witnesses to swear to tell the Truth many times each day though. It is almost robotic. There is not really much more to say about him except for that. God bless him.

Atheist 1, Acting Atheist, Atheist 2.5

There were really not just three of them. I made that up as I have a little Irish blood in my veins. They were an amalgamation of several hundred I've met since I was born. I did not embellish on their testimony though--I actually cleaned it up quite a bit to remove the profanity they tend to use in their prosecution. I had them all land in the mental hospital on purpose as I believe that is the best place to put people so blind to the world and the universe right in front them. The story about "The Moth", Charles Darwin, and the Church was the example of that. I hold no animosity towards atheists and if I did, I would be a worse hypocrite than any I perceive them to be. Their rhetoric only thickened my resolve to craft this odd book. I guess I should say "Thank You" to them for that. It is better than letting this information sit dormant in the brain cell right next to where we put the Acting Atheist.

God bless them.

The Guy on the Corner

Just a regular fella so no big whoop. He just has this funny look on his face all of the time and a shrill annoying voice. It is the stuff of the deserving of the beatings, my friends. I can promise to my God that I have never cracked him but, oh, the temptation. I will leave him alone but I'm sure the Judge, Bailiff, and the Doc would like to "talk" to him. Oh boy, would they. He is safe on my block though! God...(hard for me to say)... bless him. I can think of some other "blessings" but I'll leave it there.

Mr. Hawking

It seemed like I picked on this man the whole time. I referred to him as I believe him to be the champion of "How could a God so loving and forgiving have allowed such atrocities to happen on this earth? How could this happen to me?". For that I apologize. I was really not making fun of him. My belief is that God gave us all free-will from what seems the "lowest" to what seems the "greatest".

Mr. Hawking is a professor and a teacher. He is extremely intelligent and generates much for us to peruse. Chapter 5 was dedicated to him. As a college graduate, I know what it is like to a be a student. The professor does not require you to attend class and you are free to do whatever you want. Heck, you could go to the beach everyday. That professor will see that you don't pass the class though. He is providing you information that you need and you can choose to ignore it. Maybe we are not cut out for math and physics and we all have our limitations. Nobody is perfect. I believe that life is the same way. You only attend school for a short time but then you are turned loose to go apply it. You only attend life for a short time as well. Then? Oh, and it does not matter what our mental and physical state happens to be. We all have that line in our brain that knows the difference between right and wrong. The excuse that the Nazis used of "We were only following orders." when they would burn a building then shoot anything man, woman, or child that tried to escape, did not wash then. It did not work here on earth and it sure did not work with Mr. Hawking or God. Mr. Hawking must already understand. He would not give anyone an "A" for going to the beach with a machine gun.

God bless him.

The Jury

YOU!

God

All I can say is that I hope I did right by our God. There was much controversial information that you were given. As with "The Parrot" ghost story, it was something else forcing my hand. If this was all blasphemy or heresy then I will find out one day as will we all. None intended and God's will be done.

Amen

Printed in the United States
By Bookmasters